FLOWER BEETLES

2 0 0 9

PHOTO GRAPH

BOOK SERIES
OF THE WORLD
INSECTS

── 세계 꽃무지 컬렉션 ──

The Collection of Flower Beetles in the world

|주| 커뮤니케이션 열림

다원의 '딱정벌레 모으기' 열정
인류와 자연 진화 과정의 비밀을 밝힌 영국의 생물학자 찰스 로버트 다윈(1809~1882)
다윈은 어린 시절부터 조개껍데기, 새알, 암석과 광물, 곤충 등을 열심히 수집하러 다녔다.
특히 딱정벌레를 놓치지 않으려고 입속에까지 넣었다는 일화는 유명하다.
"살아남는 종은 강한 종도 아니고, 똑똑한 종도 아니다. 변화에 적응하는 종이다."
다윈이 남긴 말이다.

Darwin's passion for 'beetle collecting'
Britain biologist who unveiled the mystery about evolution of mankind and nature
Charles Robert Darwin (1809-1882)
Darwin collected seashells, bird's eggs, rocks and minerals, insects, etc. in his childhood.
An anecdote that he put a beetle in his mouth not to lose is famous.
"It is not the strongest of the species that survives, nor the most intelligent,
but the most responsive to change," said Darwin.

Foreword

by Son, Minwoo

꽃무지는(Cetoniid bettle)는 꽃에서 발견되기 때문에 영어로도 플라워 비틀(Flower bettle)이라고 부른다. 꽃무지는 꽃과 같이 색깔과 모양이 매우 다채로운 곤충이다. 현재까지 전 세계에는 약 3,200종의 꽃무지(Cetoniid bettles)가 알려져 있으며, 극지를 제외한 전 세계에 골고루 분포되어 있지만, 특히 아프리카와 동남아시아의 열대지역과 아열대지역에 약 70%가 서식하고 있다.

전 세계에 알려진 곤충은 약 100만~110만 종이 알려져 있으며, 해마다 새로운 종이 발견되고 있다. 그 중 꽃무지(Flower bettle)가 속해 있는 딱정벌레목 곤충은 가장 큰 분류군을 가지고 있으며, 약 40만종을 차지할 정도로 그 종류가 매우 많고 다양하다.

딱정벌레목의 특징은 두 쌍의 날개 중 초시(初翅, elytron)라 부르는 한 쌍의 앞날개는 딱딱하게 굳어 있어 갑옷을 입은 것과 같이 몸을 보호할 수 있으며, 다른 한쌍의 뒷날개는 얇은 막 상으로 앞날개보다 크며, 비행시 사용하였다가 정지시에는 앞날개 밑으로 접어 넣어 비행날개를 보호할 수 있는 보다 진보된 기능을 갖고 있다. 바로 이러한 점이 세계 각지의 고산이나 평야, 하천과 늪, 지상이나 동굴, 식물체의 내부와 외부, 흙 속 등, 거의 모든 지역에서 적응하고 다양한 종으로 번성할 수 있게 되었다고 본다.

그 중 꽃무지는 딱정벌레목(Coleoptera) 풍뎅이과(Scarabaeidae) 꽃무지아과(Cetoniinae)에 속하며, 알-애벌레-번데기-어른벌레로 4단계의 성장과정을 거치는 완전변태 곤충으로 수명은 보통 년이다.

Cetcniid bettle is called the Flower bettle in English as they are found on flowers. Like a variety of flowers, the Cetoniid bettle comes in various shapes and body colors. Approximately 3,200 species of Cetoniinae have been discovered all around the world except the north and south poles. Though their distribution is rather even worldwide, about 70% of them inhabit the African and Southeast Asian tropical and subtropical regions.

Currently, approximately 1 million to 1.1 million insect species are known worldwide with some new species discovered year by year. Coleoptera order, which includes Cetoniinae (also called flower beetle), is one of the biggest insect groups of them all with some 0.4 million species falling under it.

The order of Coleoptera is characterized by the front pair of wings, called elytron, being hard and functioning as a protective mechanism like a suit of armor while the rear pair, thin like films, is bigger than the other pair. The latter is used only when an insect in this order flies and kept under the front pair of wings for protection when it is not flying, which shows an advanced evolutionary function. This is seen as a reason this order can adapt to and diversify itself successfully in almost every ecosystem worldwide alpine area, plain, stream, wetland, cavern, inside and outside of plants, earth, etc.

Of the insect systematic categories, the Flower beetle is included in the order of Coleoptera, Scarabaeoidae family of subfamily of Cetoniinae. They go through a 4-step complete metamorphic process (egg - larva - pupa - adult) and their lifespan is about 1 year on average.

Contents

선진국의 곤충전시장에서는 심심치 않게 곤충도난사건이 벌어지곤 한다. 그만큼 선진국에서는 곤충수집이 활성화 되어 있으며 많은 사람들이 즐기는 취미이기도 하다. 지난 99년 5월 도쿄 토시마의 한 곤충점에서는 일본산 천연 사슴벌레 등 86마리가 없어졌는데 피해액이 물경 약 1억원에 달했다. 한국이라면 도저히 상상할 수 없는 일. 조그만 곤충을 수집하는 사람이나, 훔쳐가는 도둑이나, 또 이를 대대적으로 보도하는 신문 모두 생소한 일일 것이다.

우리는 이러한 생소함에 더해 학습지 성격이 강했던 곤충도감의 성격을 그래픽 모티브로서도 훌륭히 표현되어 색다른 가치를 가질 수 있음을 알리고자 한다. 이런 작은 노력이 얼마나 빛을 보게 될지는 두고봐야 겠지만...

There are occasionally burglaries of insects in an exhibition hall of insects in advanced countries. This indicates that insect collecting is many people's favorite hobby in advanced countries. 86 insects including Japanese stag beetle were stolen at an insect store in Toshima, Tokyo in May, 1999. The amount of damage for the insects was about 100 million won. It will not happen in Korea. Korean people will feel weird about a person who collects small insects, a thief who steals insects, and a newspaper which reports burglary of insects.

We would like to inform that an illustrated insect book is not just learning material, but can have unique value by great expression as graphic motive. I wish that this small effort can come to fruition.

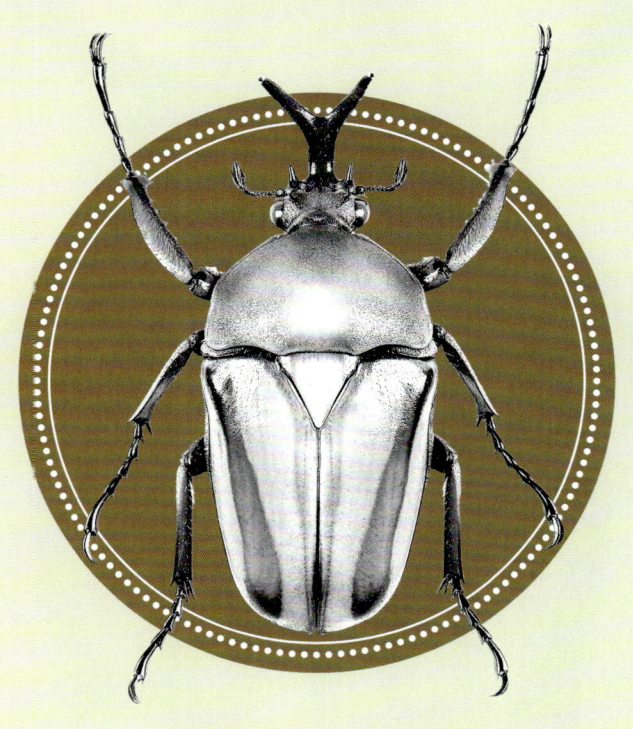

우

오스트레일리아 산 27mm

(Mt.Garmet, NQ. Australia. 2004. 9.)

Trichaulax

학 명 Scientific name	마크래이꽃무지 *Trichaulax macleayi*
채집국 Collected locality	🇦🇺 오스트레일리아 _ Australia
크 기 Size	♂♀ 26-36mm

● 채집지
Collected
Locality

● 분포지
Distribution

0°

오스트레일리아
Australia

오스트레일리아에 서식한다. 머리와 가슴은 검정색으로 광택이 있으며 딱지날개는 암적색 바탕에 황색 강모가 나있는 세 쌍의 세로줄이 있다.

They inhabit Australia. Their black cephalic part and thorax have no gloss while the dark red elytra (hard wings) have three pairs of yellow-bristled stripes.

[Distribution] Australia

♂

코트디브아르 산 91mm

(Tai Forest, Cote divoire. 2005. 7.)

Goliathus

학 명 Scientific name	골리앗레기우스대왕꽃무지 *Goliathus regius*	
채집국 Collected locality	 코트디부아르 _ Cote divoire	
크 기 Size	♂ 58-115mm, ♀ 56-82mm	

● 채집지
Collected
Locality

● 분포지
Distribution

기니   베냉   시에라리온   토고   코트디브아르   가나   나이지리아
Guinea   Benin   Sierra Leone   Togo   Cote divoire   Ghana   Nigeria

아프리카 서부에 넓게 서식한다. *G. goliatus*와 더불어 세계 최대의 무게를 가지고 있다. 살아 있을 때의 체중은 100g정도로 무겁다. 몸빛은 수컷은 광택이 없으며 암컷은 약한 광택을 지니고 있다. 머리와 가슴, 딱지날개 모두 백색바탕에 검정색의 무늬를 가진다. 가슴은 검정색을 띤 한 쌍의 반점과 두 쌍의 세로줄무늬를 가지고 있으나 가운데 한 쌍의 세로줄 무늬가 가장 크다. 넓적다리마디와 가운데가슴복판은 적갈색을 띠며 복마디는 암녹색으로 모두 광택을 지닌다.

This species, which has a wide habitat across the western part of Africa, constitutes the two heaviest insect species of the world along with *G. goliatus*. It weighs about 100g when alive. The males have no gloss while the females are slightly glossy. The cephalic part, thorax and elytra (hard wings) are white with similar black patterns, while the thorax has a pair of black speckles and two pairs of stripes, of which the inner pair of stripes has the biggest pattern. The thigh joints and mesosternum are dark-green and have a metallic gloss.

[Distribution] Sierra Leone, Guinea, Ghana, Benin, Cote divoire, Togo, Nigeria

⚥
콩고산 82mm
(Congo. 2004.)

Goliathus

학 명 Scientific name	**골리앗대왕흰꽃무지(푸스투라투스 형)** *Goliathus orientalis*
채집국 Collected locality	콩고 _ Congo
크 기 Size	♂ 55-108mm, ♀ 60-85mm

● 채집지
Collected
Locality

● 분포지
Distribution

0°

콩고
Congo

중앙아프리카 서부에 서식한다. 이 종은 *G. orientalis*의 아종으로
*G. o. preissi, G. o. undulatus*와 함께 아종을 이룬다. 딱지날개의 무늬는
*G. orientalis*와 비교하여 조밀한 망사무늬가 아닌 갈라진 땅바닥 것 같은
무늬를 가지고 있다.

This species inhabits the western part of Central Africa. Its
elytra (hard wings) pattern looks like cracked ground,
different from that of *G. orientalis*, which is mesh-like and
black. This species is divided into fore form.(orientalis,
preiss, pustulatus, undulatus)

[Distribution] Congo

♂
탄자니아산 66mm
(Tanzania. 2001. 11.)

Goliathus

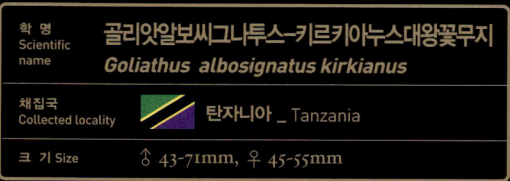

학 명 Scientific name	골리앗알보씨그나투스-키르키아누스대왕꽃무지 *Goliathus albosignatus kirkianus*
채집국 Collected locality	탄자니아 _ Tanzania
크 기 Size	♂ 43-71mm, ♀ 45-55mm

● 채집지
Collected
Locality

● 분포지
Distribution

0°

잠비아 하천유역 북부   탄자니아   말라위   짐바브웨 북부
N.region of Zambian river valleys   Tanzania   ZiMalawi   N. Zimbabwe

남아프리카 북동부에 서식한다. *Goliathus*속은 다섯 속이 있다. 이 종은 *G. a. albosignatus*와 더불어 두 개의 아종이 있으며 *Goliathus*속 중 가장 작은 아종이다. 수컷의 머리뿔은 마치 V자 처럼 발달하여 위쪽으로 굽어있다. 이 종의 몸빛은 광택이 없는 황백색을 띠며 검정색 또는 암갈색을 띤 호피무늬를 가지고 있다. 가운데종아리마디와 뒷다리 종아리마디에 나있는 강모가 황색인데 반하여, *G. a. albosignatus*는 흑색으로 이 두 아종의 구분하는데 결정적 요소가 된다.

This species inhabits the northeastern part of South Africa. The *Goliathus* genus is divided into five genera. This species, the smallest of the *Goliathus* genus, has two subgenera together with *G. a. albosignatus*. The V-shaped cephalic horns of the males are bent upward. This species' body is yellowish white with no gloss. Its yellow body with a black or dark brown tiger-skin pattern is what separates it from *G. a. albosignatus*, another subspecies which has a black body.

[Distribution] Tanzania, Northern region of Zambian river valleys, Malawi, Northern Zimbabwe

♂
코트디부아르산 79mm
(Akupe, Cote divoire(Ivory Coast). 2005. 9.)

Goliathus

학 명 Scientific name	골리앗카시쿠스대왕꽃무지 *Goliathus cacicus*
채집국 Collected locality	코트디부아르 _ Cote divoire
크 기 Size	♂ 56-98mm, ♀ 58-79mm

● 채집지
Collected
Locality

● 분포지
Distribution

0°

라이베리아   기니   코트디부아르   가나
Liberia   Guinea   Cote divoire(Ivory coast)   Ghana

아프리카 중서부에 서식한다. 이 종의 몸빛은 수컷의 경우 머리와 가슴은 황색바탕에 세 쌍의 검정색 세로줄 무늬를 가지고 있다. 딱지 날개는 회색으로 상부와 하부 양쪽으로 각각 검정색의 반점을 가지고 있으며 상부는 크고, 하부의 것은 작다. 수컷의 머리뿔은 마치 V자 처럼 발달하여 위쪽으로 굽어있으며 개체의 크기에 따라 뿔의 크기도 달라진다. 각각의 다리와 배면은 암갈색을 띠며 광택을 지닌다. 암컷의 몸빛은 광택을 지니고 있다.

The males of this species, which inhabits the mid-western part of Africa, have a yellow cephalic part and thorax with three pairs of black stripes. Its grey elytra (hard wings) have black speckles on both the upper and lower sides, and those on the former are bigger than the ones on the latter. The males' V-shaped cephalic horns are bent upward and their size is dependent upon each individual's size. Their legs and back are dark brown and glossy. The females have a glossy body.

[Distribution] Liberia, Guinea, Cote divoire, Ghana

♂
탄자니아산 48mm
(Sanya, Tanzania. 2003. 3.)

Fornasinius

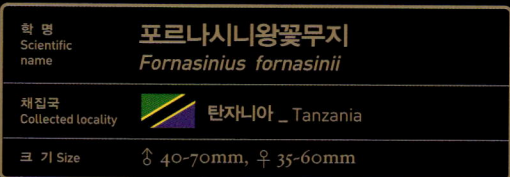

학 명 Scientific name	포르나시니왕꽃무지 *Fornasinius fornasinii*
채집국 Collected locality	탄자니아 _ Tanzania
크 기 Size	♂ 40-70mm, ♀ 35-60mm

● 채집지
Collected
Locality

● 분포지
Distribution

0°

콩그민주공화국(자이르)　모잠비크　탄자니아　르완다　부룬디　우간다 동부
D.R. Congo (Zaire)　Mozambique　Tanzania　Rwanda　Burundi　E. Uganda

중앙아프리카 남동부에 서식한다. 몸빛은 부드러운 광택을 지니며 검정색 바탕에 황색의 띠가 둘러져 있다. 가슴은 정중앙으로 한 줄의 황색 세로줄을 가지고 있으며 딱지날개는 세로줄의 황색 점무늬들이 있다. 하나의 머리뿔 말단부는 두 갈래로 뾰족하게 갈고리 모양으로 발달하였고 머리 기저부 양쪽에 가시처럼 발달한 한 쌍의 돌기가 있다.

Inhabiting southeastern part of Central Africa, they have a softly-glossy black body which is yellow-striped. On the center of the thorax runs a pair of yellow stripes while the elytra (hard wings) show some stripes of yellow speckles. Its cephalic horn is hooked and pointed, two-forked at the tip. A pair of protrusions like thorns can be seen at both sides of the lower cephalic part.

[Distribution]　Eastern Uganda, Western Kenya, Mozambique, Eastern D.R. Congo, Burundi, Rwanda, Tanzania

♂
우간다산 59mm
(Uganda. 2004. 9.)

Fornasinius

학 명 Scientific name	루쑤스왕꽃무지 *Fornasinius russus*
채집국 Collected locality	 우간다 _ Uganda
크 기 Size	♂ 40-70mm, ♀ 35-60mm

● 채집지
  Collected
  Locality

● 분포지
  Distribution

0°

가봉   콩고   콩고민주공화국(자이르)   우간다 서부
Gabon Congo   D.R. Congo(Zaire)   W. Uganda

중앙아프리카 중부에 서식한다. 몸빛은 부드러운 광택을 지니며 적갈색 바탕에 검정색의 띠가 둘러져 있다. 하나의 머리뿔 말단부는 두 갈래로 뾰족하게 갈고리 모양으로 발달하였고 *F. fornasinii*와 비교하여 머리 기저부에 양쪽으로 가시모양의 돌기가 방패형으로 더욱 크게 발달하였다.

They inhabit the middle region of Central Africa. Their dark-brown body is softly glossy and surrounded by a black stripe. Its cephalic horn is hooked and pointed, two-forked at the tip. The pointed and shield-shaped protrusions on both sides of the lower cephalic part are bigger than those of *F. fornasinii*.

[Distribution] Congo, Gabon, Northern D.R. Congo, Western Uganda

☆

타이산36mm
(Chiang-Mai, N.Thailand. 2002. 3.)

Dicranocephalus

학 명 Scientific name	왈라치사슴풍뎅이 *Dicranocephalus wallichii*
채집국 Collected locality	타이 _ Thailand
크 기 Size	♂ 24-40mm, ♀ 22-26mm

● 채집지
Collected
Locality

● 분포지
Distribution

0°

네팔　인도　부탄　미얀마·　타이　라오스　베트남　티베트
Nepal　India　Bhutan　Myanmar　Thailand　Laos　Vietnam　Tibet

서남아시아, 인도차이나 반도에 서식한다. 몸빛은 광택을 지닌 적갈색 바탕에 머리와 가슴 딱지날개는 마치 벨벳과 같이 황색의 미모로 덮여 있다. 가슴에는 한 쌍의 세로줄을 가지고 있으나 끝까지 연결되어 있지는 않다. 딱지날개 위쪽의 양쪽과 아래에 각각 한 쌍의 적갈색 반점들을 가지고 있다. 머리뿔은 기저부에서부터 두 갈래로 나뉘어져 마치 수사슴의 뿔과 닮아있다.

They inhabit the southwestern parts of Asia and the Indochinese Peninsula. Their body is glossy and reddish-brown while the cephalic part, thorax and elytra (hard wings) are covered with velvet-like fine yellow hair. There is a pair of stripes on the thorax, which gets blurry at the tips. On the upper side and lower side of the elytra (hard wings) is a pair of reddish-brown speckles. The cephalic horn, divided into two from the lower part, resembles an antler.

[Distribution] India, Nepal, Bhutan, Tibet, Myanmar, Thailand, Laos, Vietnam

♂

전북 정읍 산 27mm
(Mt. Naejangsan, Jeong-eup, Jeollabuk-do, S. Korea, 1988. 5.)

Dicarnocephalus

학 명 Scientific name	**사슴풍뎅이** ***Dicarnocephalus adamsi***
채집국 Collected locality	🇰🇷 대한민국 _ Korea
크 기 Size	♂ 23-38mm, ♀ 20-25mm

채집지
Collected
Locality

분포지
Distribution

0°

티벳   중국   한국
Tibet   China   Korea

한반도, 중국, 티벳 동부에 서식한다. 몸빛은 약한 광택을 지닌 적갈색바탕에 머리와 가슴 딱지날개는 마치 벨벳과 같이 황색의 미모로 덮여있다. 가슴에는 한 쌍의 세로줄을 가지고 있으나 끝까지 연결 되지는 않고, *D. wallichii*와 비교하여 세로줄의 띠가 더 넓고 양측에 한 쌍의 작은 반점을 가지고 있다. 딱지날개 위쪽의 양쪽과 아래에 각각 한 쌍의 적갈색 반점들을 가지고 있다. 머리뿔은 기저부에서부터 두 갈래로 나뉘어져 마치 수사슴의 뿔과 닮아있다. 서식지와 개체의 크기에 따라 머리뿔의 모양에 변화가 많다. 성충은 5월~6월초에 가장 많이 출현하며 그룹을 이루어 짝짓기 하는 모습을 관찰할 수 있다.

They inhabit the Korean Peninsula, China and the eastern part of Tibet. Their body is slightly glossy and reddish-brown while the cephalic part, thorax and elytra (hard wings) are covered with velvet-like fine yellow hair. There is a pair of stripes on the thorax, which gets blurry at the tips. The stripes, having a pair of small speckles on both sides, are wider than those of *D. wallichii*. On each of both upper sides and lower side of the elytra (hard wings) is a pair of reddish-brown speckles. The cephalic horn, divided into two from the lower part, resembles an antler. The shape of the cephalic horns depends heavily on where the habitat is and how big a specimen is. The biggest number of adult insects appears in May to early June, mating in groups.

[Distribution] Korea, China, Tibet

탄자니아산 31mm
(Mt. Uluguru, Tanzania. 2003. 7.)

Cheirolasia

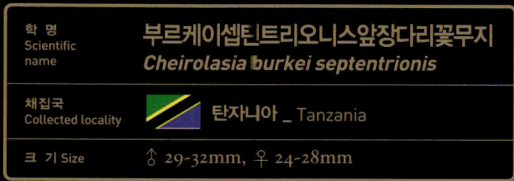

학 명 Scientific name	부르케이셉틴트리오니스앞장다리꽃무지 *Cheirolasia burkei septentrionis*
채집국 Collected locality	탄자니아 _ Tanzania
크 기 Size	♂ 29-32mm, ♀ 24-28mm

● 채집지
   Collected
   Locality

● 분포지
   Distribution

0°

콩고민주공화국(자이르)   탄자니아   케냐
D.R. Congo (Zaire)   Tanzania   Kenya

중앙아프리카에 서식한다. 이 종든 *C. b. burkei*에 비하여 밝은 노랑색을 띠고 있다. *C. b. burkei*와 함께 수컷의 앞다리는 길게 발달하여 종아리마디와 발목마디에 솔과 같은 황색의 강모를 가지고 있다. 머리뿔은 *C. b. burkei*와 비교하여 두텁다.

This species, which inhabits Central Africa, is more yellow and brighter than *C. b. burkei*. The males' forelegs, long and well-developed, have some yellow and brush-like bristles on the thigh and ankle joints like those of *C. b. burkei*. The cephalic horn is thicker than that of *C. b. burkei*.

[Distribution] D.R. Congo, Kenya, Tanzania

♂
짐바브웨산 48mm
(Guruve, Mashonaland, Zimbabwe. 2003. 5.)

Dicronorhina

학 명 Scientific name	데르비아나왕꽃무지 *Dicronorhina derbyana derbyana*
채집국 Collected locality	짐바브웨 _ Zimbabwe
크 기 Size	♂ 31-48mm, ♀ 28-44mm

● 채집지
Collected
Locality

● 분포지
Distribution

0°

콩고민주공화국(자이르)
D.R. Congo(Zaire)

앙골라
Angola   나미비아
Namibia   남아프리카공화국
Rec. of S. America   모잠비크
Mozambique   보츠나와
Botswana   탄자니아
Tanzania   잠비아
Zambia   말라위
Malawi   짐바브웨
Zimbabwe

아프리카 중남부에 넓게 서식한다. 이 종은 머리가 방패모양으로 발달하여 두 갈래로 갈라진 짧은 뿔을 가지고 있다. 몸빛은 광택을 가진 녹색 바탕에 백색의 줄무늬를 가지고 있다-. 암컷은 광택이 보다 강하다.

This species, which inhabits the mid-to-southern parts of Central Africa, has a shield-shaped cephalic part with a short, two-forked horn. Its body is green and glossy with some white stripes. The females are glossier than the males.

[Distribution] Southern Tanzania, Mozambique, Malawi, Southeastern D.R. Congo, Zambia,Rec. of South America, Angola, Botswana.

♂

탄자니아산 44mm
(E-Mt. Usambarae, Tanzania. 2004. 6.)

Dicronorhina

학 명 Scientific name	데르비아나오벨투에리왕꽃무지 *Dicronorhina derbyana oberthueri*
채집국 Collected locality	탄자니아 _ Tanzania
크 기 Size	♂ 29.5-49mm, ♀ 32-44.5mm

● 채집지
Collected
Locality

● 분포지
Distribution

0°

우간다　탄자니아　케냐　소말리아
Uganda　Tanzania　Kenya　Somalia

중앙아프리카 동부에 서식한다. 0 종은 머리가 방패모양으로 발달하여 두 갈래로 갈라진 짧은 뿔을 가지고 있다. 몸빛은 광택을 가진 황록색을 띠며 가슴의 중앙부는 황적색을 띤다. 암컷은 광택이 보다 강하다.

This species, which inhabits the eastern part of Central Africa, has a shield-shaped cephalic part with a short, two-forked horn. Its body is yellowish-green and glossy while the center of the thorax is yellowish-red. The females are glossier than the males.

[Distribution] Somalia, Uganda, Kenya, Tanzania

♂

콩고산 35mm
(PK. Rouge, Congo. 2003. 4)

Eudicella

학 명 Scientific name	그랄리뿔꽃무지 *Eudicella gralli gralli*
채집국 Collected locality	콩고 _ Congo
크 기 Size	♂ 32-44mm, ♀ 28-33mm

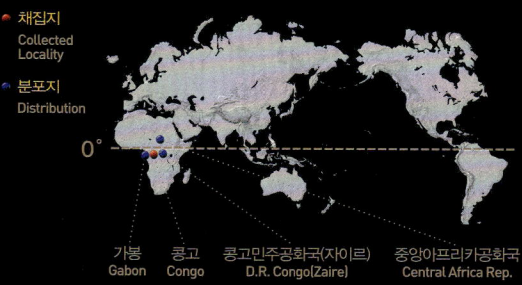

● 채집지
Collected
Locality

● 분포지
Distribution

0°

가봉
Gabon   콩고
Congo   콩고민주공화국(자이르)
D.R. Congo(Zaire)   중앙아프리카공화국
Central Africa Rep.

중앙아프리카에 넓게 서식한다. 이 종의 머리뿔은 두 갈래로 갈라져 마치 물소뿔처럼 발달하였다. 기저부의 가시와 같은 한 쌍의 돌기는 전방을 향하고 있다. 광택을 지니며 앞다리종아리마디 안쪽에 가시와 같은 돌기를 가지고 있다.

The two-forked cephalic horn of this species, which has a wide habitat across Central Africa, resembles that of a water buffalo. A pair of thorn-like protrusions on the lower part pointss forward. It has a glossy body and some thorn-like protrusion inside the frontal calf joints.

[Distribution] Central Africa Rep., Gabon, Congo, D.R. Congo

♂
케냐산 46mm
(W. Kenya. 2003. 5)

Eudicella

학 명 Scientific name	그랄리움부로비타타뿔꽃무지 *Eudicella gralli umbrovittata*
채집국 Collected locality	케냐 _ Kenya
크 기 Size	♂ 31-53mm, ♀ 27-34mm

● 채집지
Collected
Locality

● 분포지
Distribution

0°

우간다 케냐
Uganda Kenya

중앙아프리카 동부에 서식한다. 이 종의 머리뿔은 *E. g. gralli*와 비교하여 보다 크고 길게 발달하여 기저부의 가시와 같은 한 쌍의 돌기는 45° 각도로 벌어져 있다. 광택을 지니며 앞다리종아리마디 안쪽에 톱날과 같은 돌기를 가지고 있다.

This species, which inhabits the eastern part of Central Africa, has a cephalic horn bigger and longer than that of *E. g. gralli*. A pair of thorn-like protrusions on the lower part forms an oblique angle. It has a glossy body and some thorn-like protrusion inside the frontal thigh joints.

[Distribution] Uganda , Kenya

말라위산 39mm
(Malawi. 2003. 5.)

Eudicella

학 명 Scientific name	트릴리네아뿔꽃무지 *Eudicella trilineata*		
채집국 Collected locality	🇲🇼 말라위 _ Malawi		
크 기 Size	♂ 26-37mm, ♀ 25-33mm		

● 채집지
Collected
Locality

● 분포지
Distribution

0°

말라위
Malawi　　탄자니아 남부
S. Tanzania

중앙아프리카중동부에 서식한다. 이 종의 *E. s. shiratica*와 비교하여 머리 뿔은 작고 두 갈래로 갈라져 안쪽으로 굽어있다. 기저부의 돌기는 방패모양으로 발달하였다. 광택을 지니며 앞다리종아리마디 안쪽에 가시와 같은 돌기를 가지고 있다. 딱지날개 위아래 양쪽에 검은 반점을 가지고 있다.

The two-forked and inward-bent cephalic horn of this species, which inhabits the mid-to-eastern parts of Central Africa, is smaller than that of *E. s. shiratica*. The protrusion on the lower part resembles a shield. It has a glossy body and some thorn-like protrusion inside the frontal thigh joints. Some black speckles are visible underneath and on both sides of the elytra (hard wings).

[Distribution] Southern Tanzania, Malawi

♂

인도 남부 산 30mm
(Nilgiri Hills, India. 2004. 11.)

Cyphonocephalus

학 명 Scientific name	오리바세우스사슴뿔꽃무지 *Cyphonocephalus olivaceus*
채집국 Collected locality	🇮🇳 인도 _ ndia
크 기 Size	♂ 30-34mm, ♀ 22-26mm

● 채집지
Collected
Locality

● 분포지
Distribution

0°

인도 남부
S. India

인디아 남북부에 서식한다. 이 종은 1속 1종이다. 몸빛은 광택을 지닌 밝은 녹색으로 머리뿔과 종아리마디 아래로는 적갈색이다. 머리의 뿔은 마치 젊은 사슴의 뿔을 닮았다.

This species, which inhabits the northern and southern regions of India, constitutes a genus alone. It has a bright-green, glossy body and the cephalic horn is reddish-brown like the lower part of the thigh joints. The cephalic horn resembles the antler of a young deer

[Distribution] Southern India

♂
타이 북부 산 52mm
(Chiang-Mai, N. Thailand. 2003. 7.)

Jumnos

학 명 Scientific name	룩케리노랑네점박이앞장다리꽃무지 *Jumnos ruckeri*
채집국 Collected locality	타이 _ Thailand
크 기 Size	♂ 38-56mm, ♀ 38-46mm

● 채집지
Collected
Locality

● 분포지
Distribution

0°

인도동북부   미얀마   타이
E.N. India   Myanimar   Thailand

인도차이나 반도 북부에 서식한다. 이 종은 *J. r. ruckeri, J. r. tonkinensis, J. r. pfanneri*와 함께 4아종으로 나뉜다. 이 종은 녹색의 금속성 광택을 띤다. 딱지날개는 황색의 네 개의 큰 반점을 가지고 있다. *J. r. pfanneri* 만이 황색의 반점이 없거나 매우 작다.

This species, which has some metallic green gloss, inhabits the northern region of the Indochinese Peninsula. This species, with *J. r. ruckeri*, *J. r. tonkinensis* and *J. r. pfanneri*, is grouped into four subspecies. *J. r. pfanneri* has very small or no yellow speckles on its elytra (hard wings) while four yellow and big speckles are seen on those of the others.

[Distribution] Northastern India, Myanimar, Thailand

♂
카메룬산 60mm
(Mt. Cameroun, Cameroon. 2005. 2.)

Mecynorhina

학 명 Scientific name	**크라치지줄무늬귀신꽃무지** *Mecynorhina kraatzi*
채집국 Collected locality	🇨🇲 **카메룬** _ Cameroon
크 기 Size	♂ 38-70mm, ♀ 40-52mm

● 채집지
Collected
Locality

● 분포지
Distribution

0°

나이지리아
Nigeria   카메룬
Cameroon

아프리카 중서부에 서식한다. *Mecynorhina*속의 국명은 2002년도에 발간된 "세계 곤충도감"에서 저자 홍승표씨로 부터 귀신꽃무지로 명명 되어졌다. 이 종의 머리뿔은 크고 길게 발달하였으며 말단부는 두 갈래로 갈라져 있다. 머리 기저부의 가시와 같은 돌기는 작고 45˚ 각도로 벌어져 있다. 앞다리종아리마디에 돌기가 큰 톱날처럼 발달하였다.

This species inhabits the midwestern part of Africa. The local name of the *Mecynorhina* genus *Mecynorhina torquata* was first used in 2002 by Seung-Pyo Hong, when Insects of the World was published. This species has a long and well-developed cephalic horn, which is two-forked at the tip. A pair of thorn-like protrusions on the lower part forms an oblique angle. Some protrusions are well-developed like saw blades on the frontal thigh joints.

[Distribution] Nigeria, Cameroon

♂

탄자니아 산 68mm
(Mt. Zengia, Tanzania. 2005. 3.)

Mecynorhina

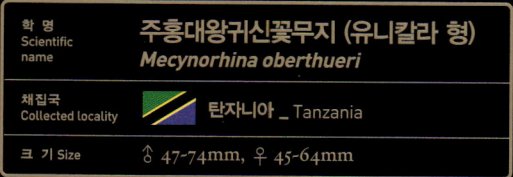

학 명 Scientific name	주홍대왕귀신꽃무지 (유니칼라 형) *Mecynorhina oberthueri*
채집국 Collected locality	탄자니아 _ Tanzania
크 기 Size	♂ 47-74mm, ♀ 45-64mm

● 채집지
Collected
Locality

● 분포지
Distribution

0°

탄자니아 북동부
NE. Tanzania

중앙아프리카 동부에 서식한다. 이 종의 머리뿔은 머리방패와 함께 발달하였고 말단부는 갈라져 있지 않다. 머리 기저부의 방패와 같은 돌기는 전방향으로 발달하였다. 앞다리종아리마디에 난 돌기는 큰 톱날처럼 발달하였다.

The cephalic horn of this species, which inhabits the eastern region of Central Africa, is developed with its cephalic shield and not forked in the tip. The shield-like protrusion on the lower cephalic part is pointed forward. Some protrusions are well-developed like saw blades on the frontal thigh joints.

[Distribution] Northestern Tanzania

♂

콩고민주공화국 산 86mm
(Khoni(Shaba), D. R. Congo. 2006. 1.)

Mecynorhina

학 명 Scientific name	토르콰타~포게이대왕귀신꽃무지 *Mecynorhina torquata poggei*
채집국 Collected locality	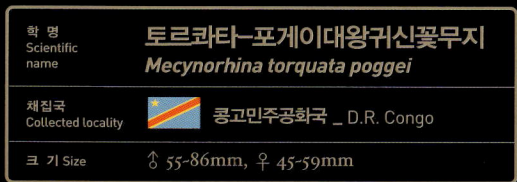 콩고민주공화국 _ D.R. Congo
크 기 Size	♂ 55-86mm, ♀ 45-59mm

● 채집지
Collected
Locality

● 분포지
Distribution

콩고민주공화국(자이르)
D.R. Congo(Zaire)

중앙아프리카 동부에 서식한다. 이종의 머리방패는 마름모형이며 뿔은 하나로 길고 크게 발달하여 위쪽으로 굽어있으며 말단부에 돌기는 *M. t. immaculicollis*와 비교하여 보다 크며 가슴과 딱지날개에 황백색의 줄무늬를 가지고 있다. 앞다리종아리마디의 돌기가 큰 톱날처럼 발달하였다.

This species inhabits the eastern part of Central Africa. The cephalic shield of the species is lozenge-shaped and the single, long and big horn is bent upward while the protrusion at the tip is bigger than that of *M. t. immaculicollis*. It has some yellowish-white stripes on the thorax and elytra (hard wings). Some protrusions are well-developed like saw blades on the frontal thigh joints.

[Distribution] Southeastern D.R. Congo

♂

콩고민주공화국 산 69mm
(N. Klvu Lake, D. R. Congo. 2006. 6.)

Mecynorhina

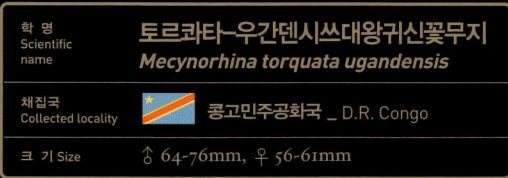

학 명 Scientific name	토르콰타-우간덴시쓰대왕귀신꽃무지 *Mecynorhina torquata ugandensis*
채집국 Collected locality	콩고민주공화국 _ D.R. Congo
크 기 Size	♂ 64-76mm, ♀ 56-61mm

● 채집지
Collected
Locality

● 분포지
Distribution

0°

콩고   콩고민주공화국(자이르)   우간다
Congo   D.R. Congo(Zaire)   Uganda

중앙아프리카에 서식한다. *M. torquata*는 *M. t. torquata, ugandensis, M. t. immaculicollis, M. t. poggei, M. t. ugandensis* 4아종으로 분류된다. 이종의 머리방패는 마름모형이며 뿔은 하나로 길고 크게 발달하여 위쪽으로 굽어있으며 말단부에 돌기를 가지고 있다. 앞다리종아리마디에 돌기가 큰 톱날처럼 발달하였다.

This species inhabits Central Africa. *M. torquata* is divided into four subspecies: *M. t. torquata*, *ugandensis*, *M. t. immaculicollis* , *M. t. poggei* and *M. t. ugandensis*,. The cephalic shield of the species is lozenge-shaped and the single, long and big horn is bent upward with some protrusions at the tip. Some protrusions are well-developed like saw blades on the frontal thigh joints.

**[Distribution] Congo, D.R. Congo, Uganda, Luanda,
Burundi**

♂
콩고민주공화국 산 76mm
(N. Klvu Lake, D. R. Congo. 2007. 6.)

Mecynorhina

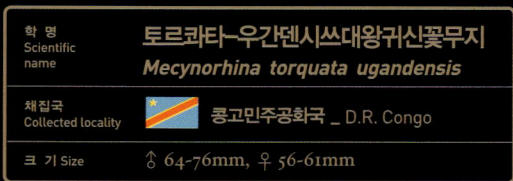

학 명 Scientific name	**토르콰타~우간덴시쓰대왕귀신꽃무지** *Mecynorhina torquata ugandensis*
채집국 Collected locality	🏴 **콩고민주공화국** _ D.R. Congo
크 기 Size	♂ 64-76mm, ♀ 56-61mm

🔴 채집지
Collected Locality

🔵 분포지
Distribution

0°

콩고
Congo   콩고민주공화국(자이르)
D.R. Congo(Zaire)   우간다
Uganda

중앙아프리카에 서식한다. 이 종은 *M. t. ugandensis*의 녹색형이다. 가슴 부위가 광택이 없는 녹색을 띤다. 이 종의 딱지날개는 선명한 적자색을 띤다. 세로줄무늬의 변이가 다양하다. *M. t. ugandensis*의 공통된 특징으로 몸빛은 광택이 없고 다리 모두는 강한 광택을 지니고 있다.

This species, inhabiting Central Africa, is the green type of *M. t. ugandensis*. Its thorax is green with no gloss while the elytra (hard wings) are reddish-purple and vivid. Its stripes show a wide range of variations. What is common for all *M. t. ugandensis* individuals is a body with no gloss and legs with some strong gloss.

[Distribution] Congo, D.R. Congo, Uganda

♂

카메룬 산 45mm
(Mt. Cameroun, Cameroon. 2005. 6.)

Mecynorhina

학 명 Scientific name	토르콰타~우간덴시쓰대왕귀신꽃무지 *Mecynorhina torquata ugandensis*
채집국 Collected locality	카메룬 _ Cameroon
크 기 Size	♂ 34~55mm, ♀ 32~38mm

● 채집지
Collected
Locality

● 분포지
Distribution

0°

카메룬
Cameroon　　중앙아프리카공화국
Central Africa Rep.　　우간다
Uganda

중앙아프리카에 서식한다. 이 종은 7~8아종을 가지고 있다. 이 아종들은 몸빛의 변이와 머리뿔의 변이가 개체별로 다르다. 그 중 *M. h. eximia*는 특히 몸빛의 변이가 다양하다. 이 종은 카메룬산으로 황색을 띠고 있다.

This species, which inhabits Central Africa, is divided into 7~8 subspecies, which show individual difference in their body colors and cephalic-horn shapes. Especially, *M. h. eximia* has a wide range of body-color variations. Native to Cameroon, this species has a yellow body.

[Distribution] Central Africa Rep., Cameroon, Uganda

♂

수마트라 서부 산 19.5mm

(W. Sumatra, Indonesia. 2002. 6)

Mystroceros

학 명 Scientific name	**로우예리쌍뿔꽃무지** *Mystroceros rouyeri*
채집국 Collected locality	인도네시아 _ Indonesia
크 기 Size	♂♀ 18.4-22.1mm

- 채집지
Collected Locality
- 분포지
Distribution

0°

수마트라 섬
Sumatera I.　자바 섬
Java I.

인도네시아 수마트라와 자바 섬에 서식한다. 이 종은 *M. macleayi*와 *M. rouyeri* 두 종이 있다. 몸빛은 암, 수 모두 강한 광택의 검정색을 띠고 있으며 녹색, 적색, 보라색의 변이가 있다. 머리뿔은 기저부에서부터 두 갈래로 나뉘어져 발달하였으며 머리방패에 두 개의 작은 돌기가 있다.

This species, consisting of *M. macleayi* and *M. rouyeri*, inhabits Sumatra Island and Java Island, Indonesia. Both males and females have a black body with strong gloss while green, red and violet variations have been discovered. The cephalic horn is two-forked from the lower part and the cephalic shield has two small protrusions.

[Distribution] Sumatera Island, Java Island

♂
중국 사천성 산 25mm
(Mt. Jinfo-shan, Sichuan, China. 2004. 6.)

Neophaedimus

학 명 Scientific name	**아우조욱시뿔꽃무지** *Neophaedimus auzouxi*
채집국 Collected locality	🇨🇳 **중국** _ China
크 기 Size	♂ 29-32mm, ♀ 24-27mm

● 채집지
Collected
Locality

● 분포지
Distribution

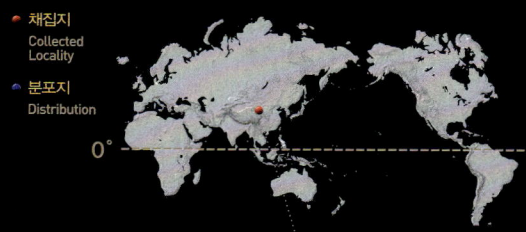

0°

중국(사천성, 운남성, 산시성, 감서성)
China(Sichuan Prov, Yunnan Prov, Shaanxi Prov, Guansu Prov)

중국 대륙에 서식한다. 이종의 몸빛은 부드러운 광택의 적갈색으로 가슴 양쪽에 한 쌍의 검은 반점과 두터운 세로줄을 가지고 있다. 배면은 전체가 검정색이다. 머리뿔은 머리방패와 함께 발달하여 말단부는 두 갈래로 갈라져 안쪽으로 굽어있으며 가슴뿔이 뾰족하게 발달하여 전방향으로 향하고 있다.

This species, which inhabits the mainland of China, has a softly-glossy and reddish-brown body with a pair of black speckles and thick stripes on both sides of the thorax while the back is entirely black. The cephalic horn, developed with the cephalic shield, is divided into two at the tip and bent inward while the pointed pronotum horn points forward.

[Distribution] China(Sichuan Prov. Yunnan Prov. Shaanxi Prov. Guansu Prov.)

♂

탄자니아 산 23mm
(Uluguru Mts., Tanzania. 2002. 4.)

Mecynorhina

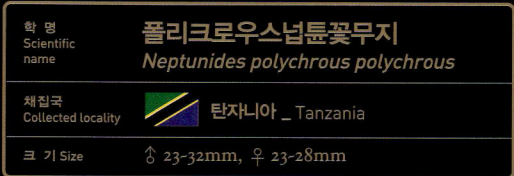

학 명 Scientific name	**폴리크로우스넙튠꽃무지** *Neptunides polychrous polychrous*
채집국 Collected locality	탄자니아 _ Tanzania
크 기 Size	♂ 23-32mm, ♀ 23-28mm

● 채집지
Collected Locality

↰ 분포지
Distribution

0°

탄자니아 동부
E. Tanzania

아프리카 탄자니아 동부에 서식한다. 이 종은 5아종이 있다. 몸빛은 강한 광택을 지니며 다양하다. 이 종은 머리와 딱지날개는 녹색, 가슴은 검정색을 띤다. 머리뿔은 머리방패와 함께 발달되어 마치 왕관모양을 하고 있다.

Inhabiting the eastern part of Tanzania, this species is divided into five subspecies. Its body, highly glossy, shows colorful variations. This species has a green and yellow body with the thorax speckled and patterned in dark brown. The cephalic horn, developed with the cephalic shield, resembles a crown.

[Distribution] Eastern Tanzania

♂

탄자니아 산 31mm

(Uluguru Mts., Tanzania. 2002. 4.)

Neptunides

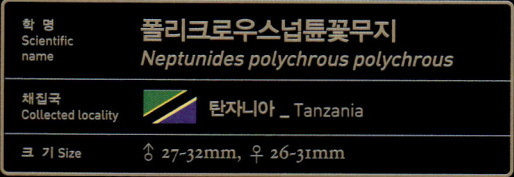

학 명 Scientific name	**폴리크로우스넵튠꽃무지** *Neptunides polychrous polychrous*
채집국 Collected locality	탄자니아 _ Tanzania
크 기 Size	♂ 27-32mm, ♀ 26-31mm

● 채집지
Collected
Locality

● 분포지
Distribution

0°

탄자니아 동부
E. Tanzania

아프리카 탄자니아 동부에 서식한다. 이 종은 50아종이 있다. 몸빛은 강한 광택을 지니며 다양하다. 이 종의 몸빛은 녹색바탕에 황색을 지니며 가슴에 암갈색의 반점과 무늬를 가지고 있다. 머리뿔은 머리방패와 함께 발달되어 마치 왕관모양을 하고 있다.

Inhabiting the eastern part of Tanzania, this species is divided into five subspecies. Its body, highly glossy, shows colorful variations. This species has a green and yellow body with the thorax speckled and patterned in dark brown. The cephalic horn, developed with the cephalic shield, resembles a crown.

[Distribution] Eastern Tanzania

♂
짐바브웨 산 33mm
(Guruve, Mashonaland, Zimbabwe. 2004. 1.)

Rhamphorrhina

학 명 Scientific name	스프레덴스흰큰머리꽃무지 ***Rhamphorrhina splendens***
채집국 Collected locality	짐바브웨 _ Zimbabwe
크 기 Size	♂ 24-34mm, ♀ 22-26mm

● 채집지
Collected
Locality

● 분포지
Distribution

0°

남아프리카공화국 북부   짐바브웨 남부
N. Rec of South Africa   S. Zimbabwe

남아프리카 중동부에 서식한다. 이 종은 *R. s. petersiana*와 함께 2아종이 있다. 몸빛은 녹색과 적색이 있으며 머리 중앙부와 가슴테두리, 딱지 날개는 백색이며 세로줄무늬를 가지고 있다. 머리방패는 직사각형 처럼 넓적하게 발달하였다.

Inhabiting the mid-to-eastern parts of South Africa, this species has two subspecies with *R. s. petersiana*. Its body color is green or red while the cephalic center, pronotum edge and elytra (hard wings) are white and striped. The cephalic shield is developed in a wide rectangular shape.

[Distribution] Southern Zimbabwe, Northern Rep.of South Africa

카메룬 산 29mm
(Cameroon. 2002. 10.)

Stephanorrhina

학 명 Scientific name	**구타타기린뿔꽃무지** *Stephanorrhina guttata*
채집국 Collected locality	🇨🇲 **카메룬** _ Cameroon
크 기 Size	♂ 22-29mm, ♀ 22-27mm

● 채집지
Collected
Locality

● 분포지
Distribution

0°

세네갈	기니	브르키나파소	코트디브아르	카메룬	베냉
Senegal	Guinea	Burkina Faso	Cote divoire (Ivory Cost)	Cameroon	Benin

중앙아프리카 북서부에 서식한다. *Stephanorrhina*속은 10종이 알려져 있으며 *S. guttata*는 6아종으로 분류된다. 몸빛은 광택을 지니며 녹색 바탕에 머리방패, 가슴테두리, 다리는 황적색을 띤다. 딱지날개는 흰 반점 들이 있다. 특히 소순판과 딱지날개 중앙부는 적색이 강하다.

Inhabiting the northwestern part of Central Africa, the *Stephanorrhina* genus is known to be divided into 10 species while *S. guttata* is classified into 6 subspecies. It has a glossy and green body while the cephalic shield, thorax edge and legs are yellowish-red. Some white speckles are seen on the elytra (hard wings). It is especially red at the center of the elytra (hard wings) and the scutellum.

[Distribution] Senegal, Guinea , Cote Divoire (Ivory Cost), Burkina Faso, Benin, Cameroon

♂

토고 산 31mm
(Kuma Forest, Kpalime, Togo. 2007. 9.)

Taurrhina

학 명 Scientific name	롱기셉스앞뿔꽃무지 *Taurrhina longiceps togolensis*		
채집국 Collected locality	🇹🇬	토고 _ Togo	
크 기 Size	♂ 24-32mm, ♀ 23-27mm		

● 채집지
Collected
Locality

● 분포지
Distribution

0°

토고　　나이지리아　카메룬　콩고　콩고민주공화국(자이르)　중앙아프리카공화국
Togo　　Nigeria　Cameroon　Congo　D.R. Congo(Zaire)　Central Africa Rep.

중앙아프리카 중앙부에 서식한다. *Taurrhina*속은 3종이 알려져 있으며 *Taurrhina longiceps*종은 5아종이 알려져 있다. 몸빛은 광택을 지니며, 녹색바탕에 가장자리들은 청록색을 띠고 있다. 머리방패는 움푹 파여져 있으며 기저부 중앙에 반타원형의 황색의 돌기가 솟아 있다. 배마디 중앙부에 적황색의 세로줄무늬가 있다.

Inhabiting the middle region of Central Africa, the *Taurrhina* genus is divided into three species with the *Taurrhina longiceps* species including five subspecies. Its body is glossy and green while the edge is greenish-blue. The cephalic shield is dented and some yellow, semi-oval protrusions are developed in the center of the lower part. The center of the abdominal sternites has some reddish-yellow stripes.

[Distribution] Togo, Nigeria, Central Africa Rep. , Cameroon, D.R. Congo, Congo

♂
카메룬 산 19mm
(Kupe, Cameroon. 2003. 10.)

Pedinorrhina

학 명 Scientific name	셉타노랑띠꽃무지 *Pedinorrhina septa*
채집국 Collected locality	 카메룬 _ Cameroon
크 기 Size	♂ ♀ 17-20mm

- 채집지
Collected
Locality
- 분포지
Distribution

0°

기니   카메룬   콩고민주공화국(자이르)   중앙아프리카공화국   우간다
Guinea   Cameroon   D.R. Congo(Zaire)   Central Africa Rep.   Uganda

중앙아프리카에 넓게 서식한다. *Pedinorrhina*속은 8종이 알려져 있으며 *P. septa*는 두 개의 이종을 가지고 있다. 몸빛은 약한 광택의 검정색을 띤다. 딱지날개는 황색의 띠를 가지고 있다.

Having a wide habitat across Central Africa, the *Pedinorrhina* genus is divided into eight species with *P. septa* having two subspecies. Its body is slightly glossy and black while the elytra (hard wings) have some yellow stripes.

[Distribution] Central Africa Rep. , Cameroon,
Guinea, D.R. Congo, Uganda

우

카메룬 산 29.2mm

(Mt. Cameroun, Cameroon, 2002. 10.)

Tmesorrhina

학 명 Scientific name	알페스트리-바후텐시스꽃무지 *Tmesorrhina alpestris bafutensis*
채집국 Collected locality	카메룬 _ Cameroon
크 기 Size	♂우 22-30mm

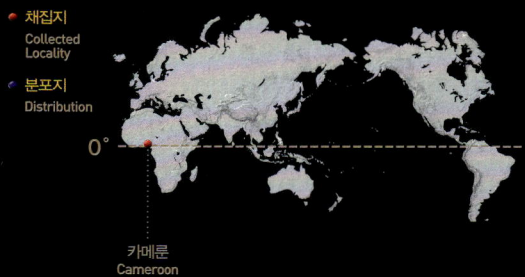

● 채집지
Collected
Locality

● 분포지
Distribution

0°

카메룬
Cameroon

중앙아프리카에 서부에 서식한다. 이 종은 *T. a. alpestris*와 비교하여 딱지날개에 암녹색의 세로줄 무늬가 있다.

This species, which inhabits the western part of Central Africa, has dark-green stripes on the elytra (hard wings) unlike *T. a. alpestris*.

[Distribution] Cameroon

♂

타이 북부 산 29mm

(Chiang - Rai, N. Thailand. 2007.10.)

Torynorrhina

학 명 Scientific name	후라메아풍이 *Torynorrhina flammea flammea*
채집국 Collected locality	🇹🇭 타이 _ Thailand
크 기 Size	♂♀ 29.5-36.5mm

● 채집지
Collected
Locality

● 분포지
Distribution

0°

| 인도
India | 미얀마
Myanmar | 말레이시아
Malaysia | 타이
Thailand | 라오스
Laos | 중국(운남)
China(Yunnan) |

인도 북부에서, 인도차이나 북부, 말레이반도, 중국 시후안성까지 서식한다. *Torynorrhina*속은 10종이 알려져 있다. *T. flammea*는 *T. chicheryi*와 더불어 두 아종으로 분류된다. 이 종은 녹색바탕의 적색형과 청색형이 있다. 다리와 배면은 청색이며 배마디는 검정색이다. 가슴배판 중앙부는 녹색을 띤다. 이 종은 농적색을 기본형으로 적색형, 녹색, 녹황색형이 있다.

Inhabiting the northern part of India, northern part of Indochina, the Malay Peninsula and Sihuan Province, China, the *Torynorrhina* genus is divided into ten species. *T. flammea* and *T. chicheryi* constitute two subspecies groups. This species is classified into green and blue variations, both of which have the body color of green. The legs and back are blue while the abdominal sternites are black. The mesosternum is green. With the dark-red variation prevailing, some red, green or yellowish-green variations have been discovered.

[Distribution] India, Laos, Myanmar, Malaysia, Thailand, China(Yunnan)

♂
서말레이시아 산 35mm
(Cameton Highland, W. Malaysia. 2008. 5.)

Torynorrhina

학 명 Scientific name	**후라메아ㅡ키케리풍이** *Torynorrhina flammea chicheryi*
채집국 Collected locality	말레이시아 _ Malaysia
크 기 Size	♂♀ 29-37mm

● 채집지
Collected
Locality

● 분포지
Distribution

0°

말레이 반도
Malay Pe.

말레이 반도에 서식한다. *Torynorrhina*속은 10종이 알려져 있다. *T. flammea*는 *T. chicheryi* 와 더불어 두 아종으로 분류된다. 이 종은 녹색형과 청색형이 있다.

Inhabiting the Malay Peninsula, the *Torynorrhina* genus is divided into ten species. *T. flammea* and *T. chicheryi* constitutes two subspecies groups. This species is classified into green and blue variations.

[Distribution] Malaysia Peninsula

♂
르완다 산 29mm
(Save, Rwanda. 2002. 5.)

Pachnoda

학 명 Scientific name	**에피피아타→활케이주홍테꽃무지** *Pachnoda ephippiata falkei*
채집국 Collected locality	르완다 _ Rwanda
크 기 Size	♂♀ 21.3-29mm

- 채집지
Collected Locality
- 분포지
Distribution

0°

르완다 우간다 케냐
Rwanda Uganda Kenya

중앙아프리카 동부에 서식한다. 이 종은 *P.e.falkei*와 *P.e.francoisi* 두 개의 이종으로 나뉘어져 있다. 이들 종은 황색바탕에 마치 로켓트와 같은 검정색의 무늬를 가지는 큰 특징이 있다. 무늬는 개체별로 변이가 있어 이들을 구분하는 데에는 어려움이 있으나 배면의 경우 *P.e.falkei*는 황색, *P.e.francoisi*는 암갈색을 띠고 있어 이 두 종을 분류하는 결정적인 요소가 된다.

This species, which is divided into two subspecies of *P.e.falkei* and *P.e.francoisi*, inhabits the eastern region of central Africa. They are characterized by some rocket-like black patterns on their yellow bodies. Though the pattern on the back, showing some individual variations, are not enough to divide the two, that of the former is yellow while that of the latter is dark brown, which facilitates their grouping.

[Distribution] Uganda, Rwanda, Kenya

우
슬로바키아 산 22mm
(Slovakia. 2004. 8)

Protaetia

학 명 Scientific name	쿠프레아점박이꽃무지 *Protaetia cuprea*
채집국 Collected locality	슬로바키아 _ Slovakia
크 기 Size	♂♀ 18.6-26.3mm

● 채집지
Collected
Locality

● 분포지
Distribution

0°

유럽(이탈리아, 터키, 아제르바이잔,헝가리, 등)
Europe (Itallia, Slovakia, Turkey, Azerbaidzhan, Hungary, etc.)   중국
China   몽골
Mongol   사할린
Sakhalin I.

유럽, 중국, 사할린에 이르기까지 구북구(Paleorctic region) 전반에 넓게 서식한다. 이 종의 몸빛은 황녹색, 금녹색, 녹색, 적동색, 녹황색 등 색체의 변이가 많고, 점무늬가 있는 것과 없는 것 등 서로 다른 종이라 할 만큼 변이가 많은 종이다. 모두가 광택을 지니고 있다.

This species has an expansive habitat in the Paleoarctic region from Europe and China to Sakhalin. Its body color shows a very wide range of individual variations from yellowish-green, golden-green, green, red-copper to greenish-yellow. So does its spotted pattern, some individuals having one unlike others. All of them are glossy.

[Distribution] Europe (Itallia, Slovakia, Turkey, Azerbaidzhan, Hungary, etc.), Sakhalin Island, Mongol, China

우
우크라이나 산 21mm
(Ukraine. 2004.)

Netocia

학 명 Scientific name	헝가리카꽃무지 *Protaetia hungarica*
채집국 Collected locality	우크라이나 _ Ukraine
크 기 Size	♂♀21mm

● 채집지
Collected
Locality

● 분포지
Distribution

0°

우크라이나
Ukraina

동유럽에 서식한다. *Netocia*속은 약 40여종이 알려져 있다. 이 종은 녹색의 금속성 광택을 지니고 있으며 가슴에 네 개의 금색 점과 딱지 날개에 반점들을 가지고 있지만 개체별로 변이가 있다.

The *Netocia* genus, which inhabits Eastern Europe, is divided into some 40 species. This species, having a metallic-green gloss, has four golden spots on its thorax and some speckles on its elytra (hard wings), though there are some individual variations.

[Distribution] Ukraina

♂

전북 부안 산 22mm

(Mt. Byeonsan, Buan-gun, Jeollabuk-do, S. Korea. 2006. 5.)

Protaetia

학 명 Scientific name	점박이꽃무지 *Protaetia orientalis submarmorea*
채집국 Collected locality	🇰🇷 대한민국 _ Korea
크 기 Size	♂♀ 16-27mm

● 채집지
Collected
Locality

● 분포지
Distribution

| 히말라야
Himalaya Mts. | 인도
India | 중국
China | 타이완
Taiwan | 대한민국
Korea | 일본
Japan |

한반도를 비롯하여 인도, 중국, 일본에 이르기까지 넓게 서식한다. 이 종의 몸빛은 황동색, 녹동색으로 금속성 광택이 있으며 배면은 더욱 강하다. *P. brevitarsis seulensis*에 비하여 몸 전체의 점각이 두드러져 있으며 배마디 양쪽으로 황백색의 반점이 있다. 딱지날개의 황백색의 점무늬는 변이가 많다.

This species has an expansive habitat from the Korean Peninsula, India, China to Japan. Its body is copper-brown or copper-green with some metallic gloss, which is more prominent on the back. Compared with *P. brevitarsis seulensis*, it shows more prominent stipples across the body, having some yellowish-white speckles on both sides of its abdominal segments. The yellowish-white spot pattern on the elytra (hard wings) shows a wide range of variations.

[Distribution] Korea, India, Himalaya Mountains, China, Taiwan, Guam, Japan

우

에콰도르 산 20mm

(Ecuador. 2005. 7.)

Gymnetis

학 명 Scientific name	남미점박이모가슴꽃무지 *Gymnetis pantherina*
채집국 Collected locality	에콰도르 _ Ecuador
크 기 Size	♂♀ 20-29mm

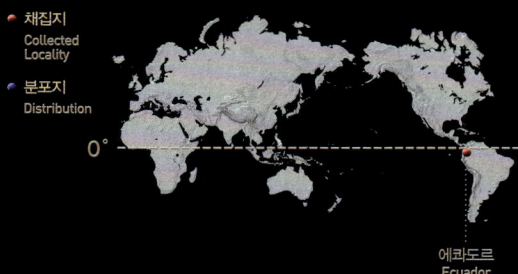

● 채집지
Collected
Locality

● 분포지
Distribution

0°

에콰도르
Ecuador

Gymnetis 속은 중앙아메리카에서 남아메리카 중부까지 넓게 서식하고 있다. 이 종의 몸빛은 마치 벨벳과 같이 부드러운 황갈색의 미모로 덮여 있으며 광택이 없는 황색바탕에 많은 검정색의 점무늬를 가지고 있다. 방사형 점무늬의 변이가 다양하다.

The *Gymnetis* genus have an expansive habitat from central Africa to the central region of South America. This species flat-yellow bodies, dotted with numerous black speckles, are covered with fine and soft yellowish-brown hair like velvet. The radial-shaped spotted pattern shows a wide range of variations.

[Distribution] Ecuador

♂

필리핀 네그로스 산 24mm
(Mt. Canlaon, N. Negros I., Philippines 2006. 5.)

Phaedimus

학 명 Scientific name	하우드니투구꽃무지 *Phaedimus howdeni*
채집국 Collected locality	 필리핀 _ Philippines
크 기 Size	♂ 22-26mm, ♀ 19-21mm

● 채집지
Collected
Locality

● 분포지
Distribution

0°

필리핀(네그로스 섬)
Philippines(Negros I.)

필리핀 네그로스 섬에 서식한다. 머리와 가슴뿔은 마치 장수풍뎅이 처럼 발달되어 있다. 몸빛은 금속성 광택의 녹색으로 딱지날개는 황등색의 넓은 띠를 가지고 있다. 머리뿔은 발달되어 안쪽으로 굽어있으며 말단부는 두 갈래로 나뉘어져 있다. 소형의 개체에서는 말단부가 갈라지지 못하여 부채모양을 하고 있다.

They inhabit Negros Island, Philippines. The cephalic horns and pronotum horns are well-developed like those of rhinoceros beetles. Their body is green and metallic-glossy while the orange-yellow elytra have some wide stripes. he cephalic horns are well-developed and bent inward while the tips are divided into two. Some small individuals` cephalic horn tips, not divided, resemble a fan.

[Distribution] Philippines(Negros Island)

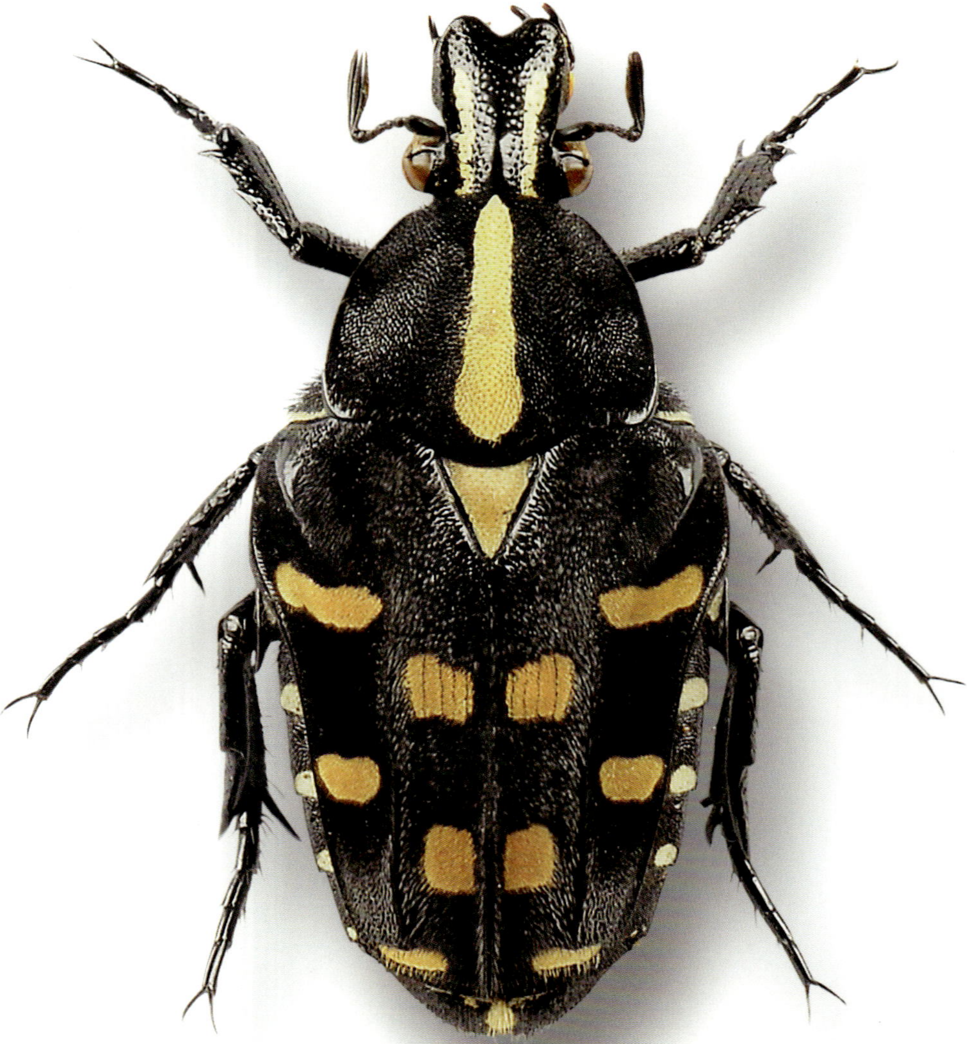

♂
술라웨시 산 22mm
(Palopo, Sulawesi, Indonesia. 2006. 3.)

Ixorida

학 명 Scientific name	엘레강스점박이홀쭉꽃무지 *Ixorida elegans*
채집국 Collected locality	 인도네시아 _ Indonesia
크 기 Size	♂♀ 17-22mm

● 채집지
Collected
Locality

● 분포지
Distribution

0°

술라웨시
Sulawesi I.

인도네시아 술라웨시 섬에 서식한다. *Ixorida*속은 62여종이 알려져 있다. 이 속 또한, 딱지날개 양쪽의 가장자리가 안쪽으로 길게 움푹 들어가 홀쭉해진 특징을 가진다. 이 종의 몸빛은 광택이 없는 검정색에 황색반점을 가지고 있으며 가슴에 황색의 세로줄을 가지고 있다.

The *Ixorida* genus, which inhabits Sulawesi Island, is divided into 62 species. This genus is characterized by its elytra (hard wings) whose slim edges reach deep inside. This species has some yellow speckles on its flat black body, while there are some yellow stripes on the thorax.

[Distribution] Sulawesi Island

♂
인도네시아 말루쿠 제도 산 16mm
(Maluku Is., Indonesia. 2006. 7.)

Ixorida

학 명 Scientific name	아펠레스주홍줄홀쭉꽃무지 *Ixorida venerea apelles*
채집국 Collected locality	인도네시아 _ Indonesia
크 기 Size	♂♀ 15-17mm

● 채집지
Collected
Locality

● 분포지
Distribution

0°

인도네시아(말루쿠 제도)
Indonesia (Maluku Is.)

인도네시아 말루쿠 제도에 서식한다. *Ixorida*속은 62여 종이 알려져 있다. 몸빛은 가슴이 검정색이고 나머지는 적갈색이며 약한 광택을 가지고 있다. 머리에 두 개의 세로줄, 가슴에 세 개의 세로줄, 딱지날개에 서로 연결되지 않은 3쌍의 세로줄 무늬는 짙은 황색의 미모로 덮여있다.

The *Ixorida* genus, which inhabits Maluku Island, is divided into 62 species. The thorax is black while the rest of the body is reddish-brown, subtly glossy. The two cephalic stripes, three on the thorax and another six, disconnected on the elytra (hard wings), are covered with fine dark-yellow hair.

[Distribution] Indonesia(Maluku Islands)

우
필리핀 민다나오 섬산 28mm
(Mindanao I., Philippines. 2007. 5.)

Plectrone

학 명 Scientific name	앤드로애디홀쪽꽃무지 *Plectrone endroedii*	
채집국 Collected locality		필리핀 _ Philippines
크 기 Size	♂♀ 24-29.3mm	

● 채집지
Collected
Locality

● 분포지
Distribution

0°

필리핀(민다나오 섬,사마이루 섬, 레이트 섬)
Philippines (Mindanao I., Samairu I., Leyte I.)

필리핀의 민다나오 섬에 서식한다. 이 종은 고광택의 금속성 색감을 지니고 있다. 녹색, 청색, 검정색 등의 색깔을 가지고 있으며 금속 세공품을 보는 듯하다.

This species, which inhabits Mindanao Island, Philippines, is highly-glossy and metallic. Its various colors (green, blue, black and so on) are like a cafted matal piece.

[Distribution] Philippines (Mindanao, Samar, Leyte Island)

동티벳 홍위엔 고원 해발 3500m
E. Tibet Plat. Alt. 3,500m. July. 24. 2007.

20여 년에 걸쳐 세계를 누비며 채집한
곤충 만큼이나 소중한 추억들을 회상한다.

병을 얻어 몸을 가눌 수 없어
비틀거릴때 마다 누울 곳을 말없이
마련해 주고 음식까지 내어주던
소박한 눈빛과 수줍은 웃음.

혹여나 허송세월하는 것이 아닌가
자책할 때면 고립된 순수를 머금은 이슬을
내뱉으며 내 눈과 가슴을 일깨워 주던
낯설은 꽃잎들.

나눌수 없는 기억들이지만 적어도
이 한켠에는 몇장의 추억들을 진열해
보고자 한다.

동티벳 홍위앤(弘原) 고원

어린이 목동 Shepherd boy

김태완소장과 제자 한상탁
Chef Kim, Taewan and his disciple Han, Sangtak collect a dung beetle from Yak's excrement

한태(설치류)가 섭취한 딱정벌레 Carabidae eaten by rodents

티벳고원의 장족 여자아이 Girl of Jang family

짐을 나르는 야크 Yak carry a burden

I look back on memories as precious as insects that I have collected all over the world for about 20 years. People with clean eyes and shy smile who provide a sleeping place and food when I was sick and staggered, Strange petals that drop lonely and pure dew on me to open my eyes and mind when I blamed myself for wasting time, I would like to display some memories on this page although I can not share all the memories.

E. Tibet Plat. Alt. 3,500m. July. 24. 2007.

Photograph Book Series of the World Insects

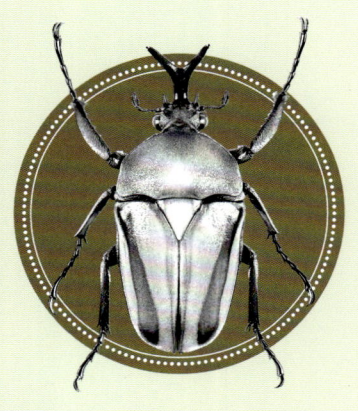

"It is not the strongest of the species that survives,
nor the most intelligent, but the most responsive to change,"
Charles Robert Darwin (1809-1882)

펴낸이
박철영
(주)커뮤니케이션 열림 경기도 파주시 교하읍 문발리 파주출판도시 514-7
TEL. 031)955-0123 FAX. 031)955-0119
www.comopen.co.kr

2009년 4월 20일 인쇄
2009년 4월 28일 발행

Published by
Park, Cheolyoung
Communication Yeollim Co., Ltd.
514-7, Paju Book City, Munbal-ri, Gyoha-eup, Paju-si, Gyeonggi-do,
SEOUL 413-756 KOREA.
TEL. +82-31-955-0123 FAX. +82-31-955-0119

First printed in 2009
First published in 2009

저자(사진, 글)
손민우

Author(Photograph & write)
Son, Minwoo

ISBN 978-89-93849-02-8-76490
ISBN 978-89-959228-9-7-76490(세트)

PHOTO GRAPH

BOOK SERIES

OF THE WORLD

INSECTS